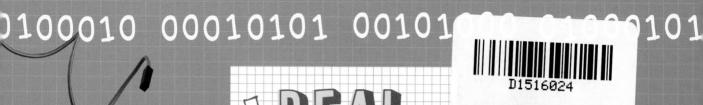

0100010 00010101 0010100 0100101

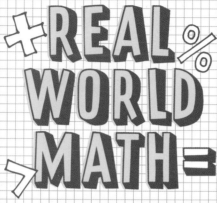

REAL WORLD MATH

1 2 3 4

CODING

by Jennifer Szymanski

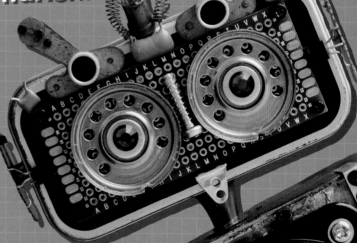

<@01>

Children's Press®
An imprint of Scholastic Inc.

Library of Congress Cataloging-in-Publication Data
Names: Szymanski, Jennifer, author.
Title: Coding / Jennifer Szymanski.
Description: First edition. | New York : Children's Press, an imprint of Scholastic Inc, 2021. | Series: Real world math | Includes index.
Audience: Ages 5–7. | Audience: Grades K–1. | Summary: "This book introduces young readers to math concepts about coding"—Provided by publisher.
Identifiers: LCCN 2021000037 (print) | LCCN 2021000038 (ebook) | ISBN 9781338761917 (library binding) | ISBN 9781338761931 (paperback) | ISBN 9781338761948 (ebook)
Subjects: LCSH: Computer programming—Juvenile literature. | Coding theory—Juvenile literature.
Classification: LCC QA76.6115 .S963 2021 (print) | LCC QA76.6115 (ebook) | DDC 005.13—dc23
LC record available at https://lccn.loc.gov/2021000037
LC ebook record available at https://lccn.loc.gov/2021000038

10 9 8 7 6 5 4 3 2 1   22 23 24 25 26

Printed in Heshan, China 62
First edition, 2022

Series produced by WonderLab Group, LLC
Book design by Moduza Design
Photo editing by Annette Kiesow
Educational consulting by Leigh Hamilton
Copyediting by Vivian Suchman
Proofreading by Molly Reid
Indexing by Connie Binder

Photos ©: 1 bottom left: Wabeno/Dreamstime; 2 bottom left: Aleksander Kovaltchuk/Dreamstime; 5 top, bottom right: Decoy Games; 6–7: Gorodok495/Dreamstime; 22 bottom: Kittipong Jirasukhanont/Dreamstime; 27 bottom: Topgeek/Dreamstime; 28–29 all: Decoy Games.

All other photos © Shutterstock.

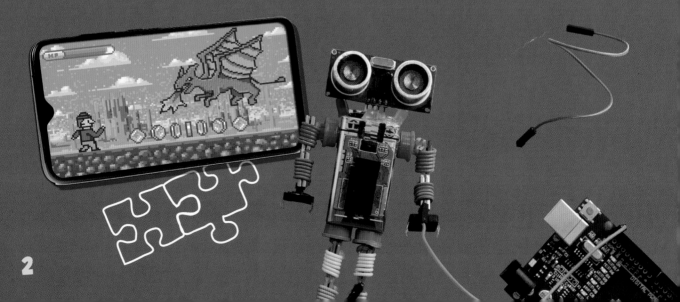

CONTENTS

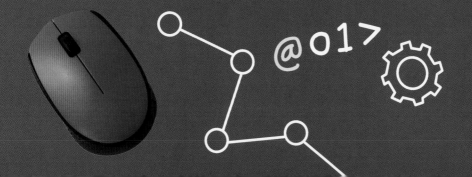

@01>

LET'S GO!

A robot dog barks. A rocket ship blasts into outer space. Your video game beeps—you've won! These and so many other things in our lives are powered by computers.

Computers wouldn't be able to do any of these things without **programmers**. Computers cannot make decisions, but they can follow instructions, called a **program**. Programmers use math to write instructions in a special way that computers can understand called **code**. Code can tell a computer to complete a task in steps. It can also be written in **patterns**. And programmers can solve a problem by breaking it down into smaller parts.

Today we are going to program a robot to help us around the house. For that, we'll need some code. Are you ready?

Ahmed and Khalil

MEET KHALIL

Khalil Abdullah is going to help us with our robot! Khalil and his brother Ahmed have been playing computer games since they were three and four years old. They loved games so much that they started their own video game company. Part of Khalil's job is to write video game code. His work is what makes their company's games so fun!

<@010>

First we need to decide what our robot will do. Remember, the robot is a computer and cannot think like a person. Computers can only follow a set of exact directions called a **program**.

1 2 3 4

CODE WITH KHALIL

Programmers write programs by using **logic**. Logic means thinking about a problem in a way that makes sense. Khalil uses logic to write the code that makes up the steps of a video game program. For example, he wouldn't write code that tells a computer to show the words "GAME OVER" first.

When we do an activity, we usually don't think too much about the steps we need to take. Think about coloring a picture. We just grab a colored pencil and get to work. But if we want the robot to color, we need to make an exact plan that lists all the steps. Then we need to make sure the steps are in the right order.

7

Whew, all of this talking about programs has made us thirsty!

Let's program the robot to get us a drink of water.

Here are the steps we need to write in our program. If the robot follows them in the wrong order, we're not going to get a drink. And the robot might make a mess.

YOU CAN DO IT! 1

Put these steps in the right order to program the robot.

> **Turn on the tap.**

4> **Wait for the water to fill the cup.**

> **Get a cup.**

5> **Turn off the tap.**

2> **Put the cup under the tap.**

Good job writing your first program for the robot! Now it's time to learn the language of computers.

10100010

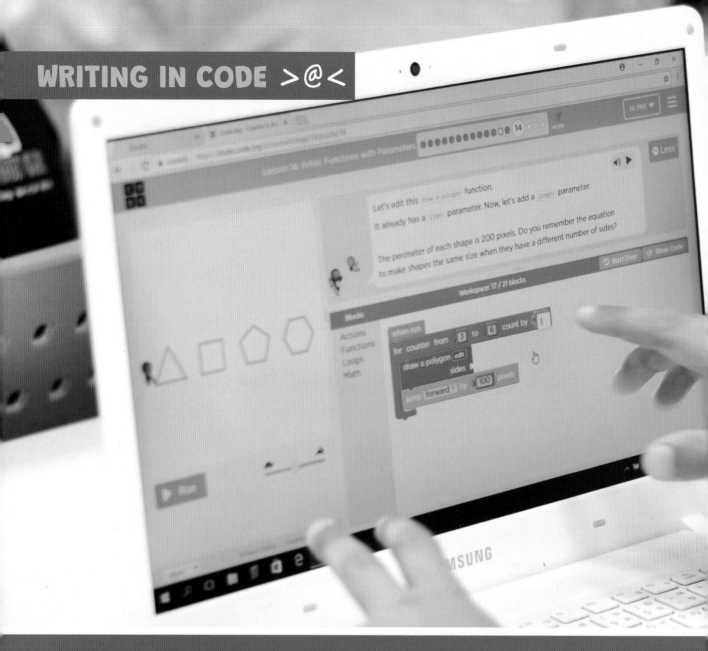

Now we have to decide how to talk to the robot. When we need to give someone directions, we say or write words. A robot cannot understand words the way people do. So programmers write computer directions in code.

<@017

`<;*1O1>`

`(*^_^*)`

`<@O1>`

There are different kinds of code. Some code is written in letters and numbers. Some uses symbols like * or /. But all code has the same job. It turns a programmer's ideas into directions a computer can follow.

`<@>`

WALK THE DOG

How can we write this phrase in code? The code we're going to use has a **key**. You can see the key at the bottom of this page. In the key, every letter of the alphabet is matched up with a number. Find each letter in the phrase on the key. The number that matches the letter is the number we should write in the code. **WALK THE DOG** in this code would be:

23 1 12 11   20 8 5   4 15 7
‾‾ ‾ ‾‾ ‾‾   ‾‾ ‾ ‾   ‾ ‾‾ ‾

KEY

A	B	C	D	E	F	G	H	I	J	K	L	M
1	2	3	4	5	6	7	8	9	10	11	12	13
N	O	P	Q	R	S	T	U	V	W	X	Y	Z
14	15	16	17	18	19	20	21	22	23	24	25	26

YOU CAN DO IT! ②

Computers use code to talk to each other. But not to talk to humans! Computers turn code back into words so that we can read them. The robot has a message in code.

Using the key on page twelve, can you figure out what the code says?

g e t   t h e   l e a s h
7 5 20   20 8 5   12 5 1 19 8 !

Wow, you did a great job!

Our robot is going to take a quick break while we dig a little deeper into code.

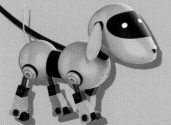

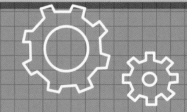

10100010

 Before a robot takes action, it changes the program's numbers and letters into a special kind of code. This type of code has two parts and is called **binary** code.

14

To write and read this kind of code,

we'll have to look for patterns. Patterns are sets of objects, like numbers or shapes, that can repeat. One type of binary code uses two numbers: the number 0 and the number 1.

10100010100

Programmers need to understand patterns in binary code before they write programs. Think about all of the colors in a video game. Every color is made by a different pattern of 1s and 0s in the code. That's a lot of patterns made out of only two numbers!

Say we have a piece of code that has two spaces. Each space can have a 0 or a 1. How many different ways can we fill the spaces?

We can fill the spaces with 00, 11, 10, or 01. That's four different ways. And the computer reads each one as a different instruction.

101000 1000

000

0_000

10100010

YOU CAN DO IT!

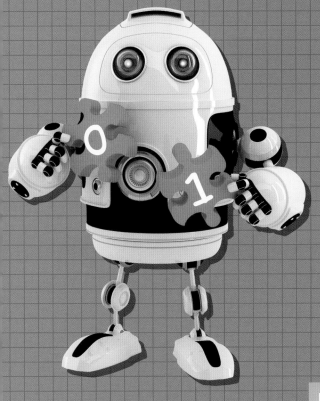

For some tasks, we have to give the robot lots of different directions. So we need to make as many patterns as we can. This time we have a piece of code that has more spaces! How many different patterns of 1 and 0 can you make?

1 __ 0 __ 1 __

Excellent—the robot is ready to go.

Just in time, because we're getting hungry.

It can be hard for a person to make a decision. But it's impossible for a computer. It needs our help. As programmers, we use code to help a robot make choices.

<@01>

18

CODE WITH KHALIL

Writing code for video games means thinking a lot about "if" and "then." When Khalil develops a video game, he asks himself, "What should happen if...?" The answer might be if the player earns 1,000 points, then they go to a new level. Or maybe if a character finds a magic coin, then they get a superpower.

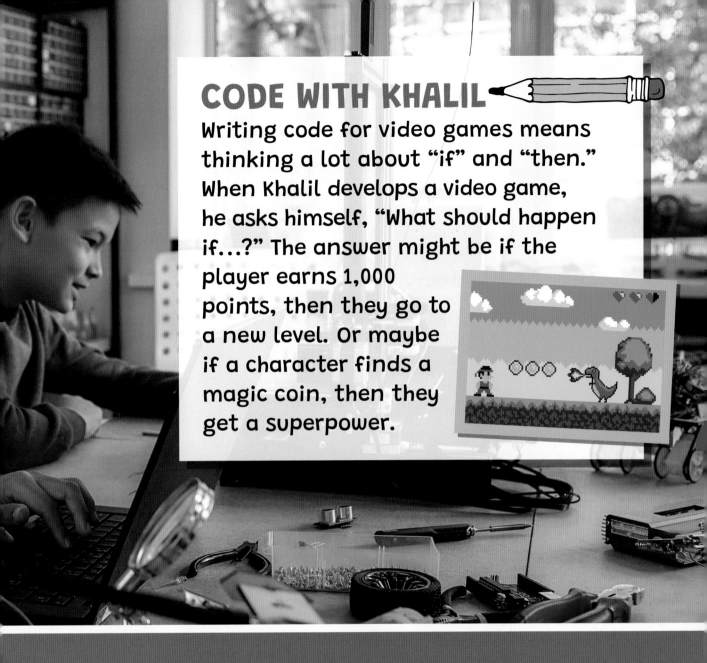

A lot of the decisions we make depend on something that happens first. If a traffic light turns green, then we go forward. If it's cold, then we put on a coat. Programmers use the same words when writing code: **if** and **then**.

Let's have the robot get us something to eat.

There are four kinds of fruit in the kitchen. We need to help the robot decide what kind of fruit to bring.

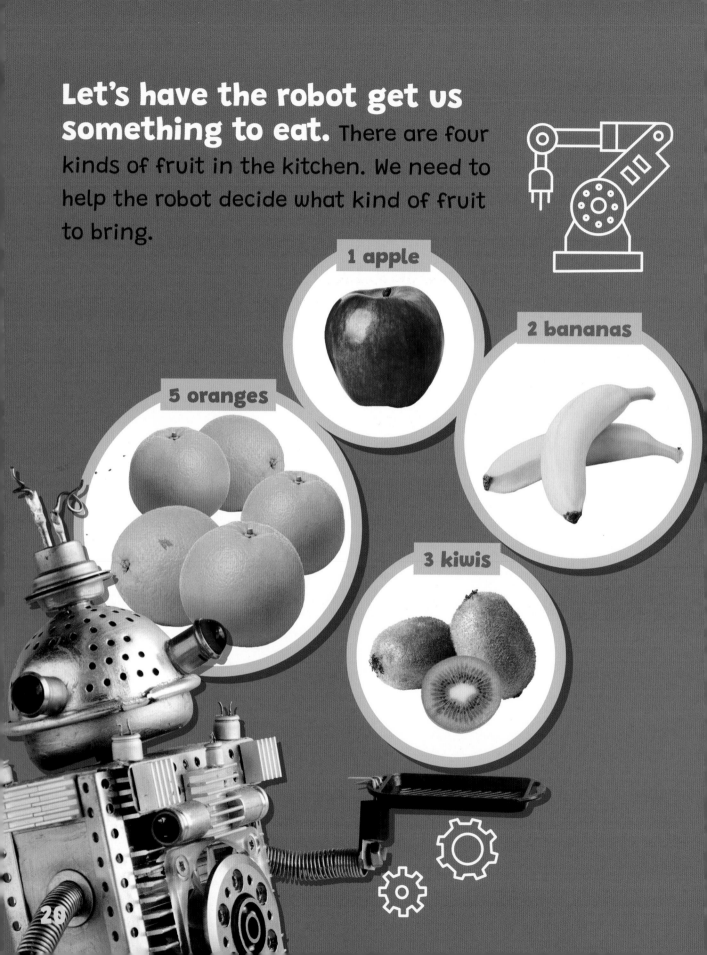

1 apple

2 bananas

5 oranges

3 kiwis

20

YOU CAN DO IT!

4

Count the number of pieces in each group of fruit. Then read the ways we can program the robot. What will happen if the robot follows each program?

If we program the robot to bring fruit from the group...

A ...with the greatest number, **then** it will bring ___oranges___.

B ...with the smallest number, **then** it will bring ___apple___.

C ...that has a number greater than 3, **then** it will bring ___oranges___.

Hooray! We've gotten our snack. Now it's time to take on the biggest job of all ... a messy bedroom!

00101000

If you've ever had to clean up a messy room, you know it can be hard. Where should you start? It helps to pick one smaller job to do first. Maybe you put your clothes away, then make your bed.

Problems can seem big to programmers, too. So they break them into smaller parts. They solve each small problem, then put the answers together to solve the big one. It's like working on a jigsaw puzzle. When you put the pieces together, you can see the whole picture.

`<;@*>`
`<101`

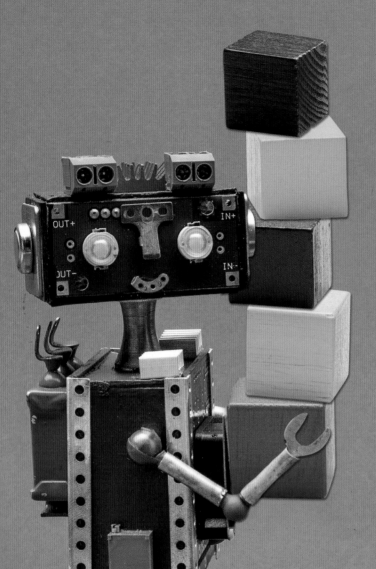

Ready to break it down?

We want to program the robot to pick up ten blocks in a room. But there's a problem: It cannot carry all ten at once. It needs to make two trips. Think about how to break down ten into two parts. It might pick up five toys on the first trip, and five on the second trip.

<;*101>

<@01>

YOU CAN DO IT! 5

How else can you break down ten into two parts? They don't have to be equal. How many blocks would the robot need to carry on each trip?

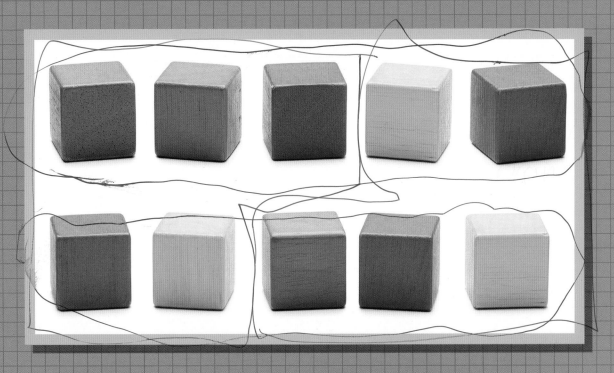

Good job! The room is neat and tidy. It's a perfect place to relax after all the coding we've done!

1010001010

WAY TO GO!

Computers are almost everywhere. They power our video games and our phones. They make cars safer and help drivers know where to go. Computers help doctors treat sick or injured people. And they help connect people all over the world.

a doctor with a patient

Programmers use math skills every day to write code. They solve problems to make the computers we use better than ever. Math is so important every day and in so many ways. You might be surprised how often you use it!

YOU CAN DO IT!

What else can you write in code? Use the key on page twelve to write out other instructions or even your name! Try making up your own code. How would you use it?

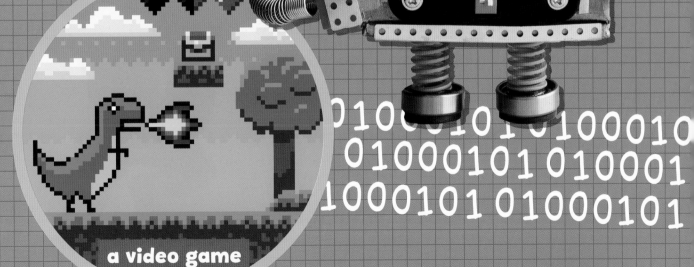

a video game

KHALIL CODES

Khalil Abdullah is a video game developer. He founded and owns a video game company, Decoy Games, with his brother Ahmed. Video game developers design and produce video games. They design characters, add sound effects, test the game, and so much more. Game developers also do lots of programming. They write the code needed for the games and apps that people enjoy.

Writing computer code involves math skills. These skills don't

010
000
100

01000101000

always look like the numbers you might see in math class. Instead, coding is a lot of problem-solving. Video game developers use the skills they learned in math class, plus their creativity, to design a game from start to finish.

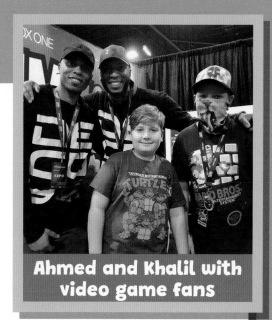
Ahmed and Khalil with video game fans

It takes imagination to look at a problem in a new way, and logic to think about how to solve a problem in a way that makes sense. Both of these things are used in math. And they're also used in the amazing video games developed by Khalil and his company!

Khalil with fans at a gaming convention

1010001010000 !

GLOSSARY

binary (BYE-ner-ee): files and code that convert all numbers and letters into strings of 1s and 0s

code (kode): the instructions that make up a computer program, written in a programming language

if/then (if/THEN): a relationship in which one thing causes something else to happen

key (kee): something that provides a solution or an explanation

logic (LAH-jik): good or valid thinking or reasoning, as in a well-written code with solid logic

pattern (PAT-urn): a repeating arrangement of colors, shapes, and figures

program (PROH-gram): a series of instructions, written in a computer language, that controls the way a computer works

programmer (PROH-gram-ur): a person whose job is to program a computer

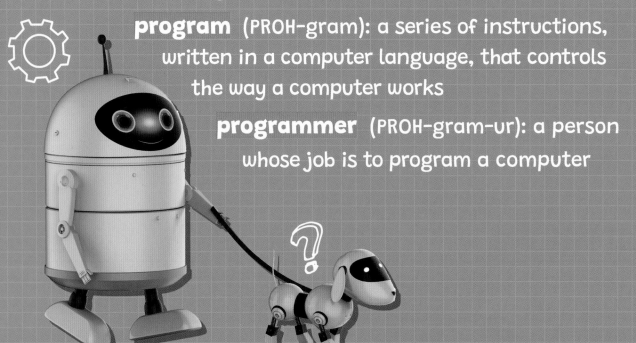

YOU CAN DO IT! ANSWER KEY

PAGE 9

1. Get a cup.
2. Turn on the tap.
3. Put the cup under the tap.
4. Wait for the water to fill the cup.
5. Turn off the tap.

PAGE 13

Get the leash!

PAGE 17

8 different patterns (000, 111, 001, 010, 011, 100, 110, 101)

PAGE 21

A. oranges; B. an apple; C. oranges

PAGE 25

1 and 9; 2 and 8; 3 and 7; 4 and 6

INDEX

Page numbers in **bold** indicate illustrations.

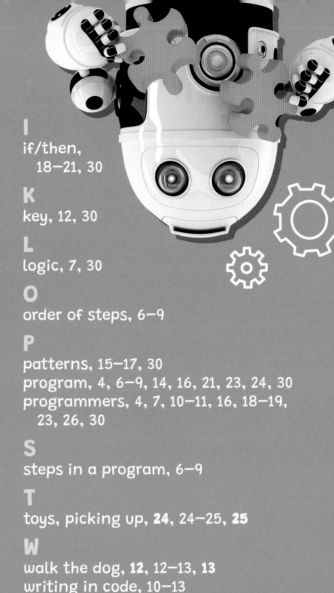

ABOUT THE AUTHOR

Jennifer Szymanski is an author and freelance science education writer. She specializes in writing materials that support both teachers and students in meeting national and state science standards but considers her "real" job to be helping students connect science to everyday life. She lives near Pittsburgh, Pennsylvania, with her family and two cats.